Handbook for Phase 1 Habitat Survey
- a technique for environmental audit

Field Manual

T0300836

England Field Unit
Nature Conservancy Council
1990

Revised reprint 2003 © JNCC 1993, 2003

This reprint edition published by Pelagic Publishing 2012
www.pelagicpublishing.com

ISBN-13   978-1-907807-24-4

This book is a reprint edition of ISBN-10   0-86139-637-5

Preface - 2003 edition

The Phase 1 habitat classification and methodology, as published in this survey handbook by the Nature Conservancy Council in 1990 (reprinted by JNCC in 1993), has been widely used throughout Britain for a diverse range of purposes. It has largely stood the test of time, and continues to be used as the standard 'phase 1' technique for habitat survey across the UK.

The responsibilities of the former Nature Conservancy Council are now held by the three country agencies, English Nature, Scottish Natural Heritage, Countryside Council for Wales and the Environment and Heritage Service, Northern Ireland with the JNCC maintaining common standards across Britain. Queries relating to this handbook should be directed to the Habitats Team of JNCC.

No revisions have been made to the handbook at this stage (except for Appendix 8), though it is intended that a full review will be undertaken at a later date to incorporate the wealth of experience gained from its use over the years, to reflect relevant technical developments, and incorporate other changes. The Operational guidelines (Part 1) remain largely relevant, although a few sections are rather out of date. The key points can be found in Appendix 9 on page 62.

Contents

		Page
1	**Introduction**	1
2	**Habitat definitions**	2
	A Woodland and scrub	2
	B Grassland and marsh	4
	C Tall herb and fern	9
	D Heathland	9
	E Mire	11
	F Swamp, marginal and inundation	15
	G Open water	17
	H Coastland	20
	I Rock exposure and waste	22
	J Miscellaneous	24

Appendices

1	Phase 1 survey habitat classification, hierarchial alphanumeric reference codes and mapping colour codes	27
2	Habitat codes for use on monochrome field maps and fair maps	38
3	Dominant species codes	42
4	Key words and status categories for target notes	46
5	Hypothetical examples of target notes	51
6	Standard recording forms	53
7	The NCC/RSNC habitat classification	56
8	Relationship between Phase 1 habitat categories and National Vegetation Classification communities	60
9	Technical developments and other changes since 1990 - key points	62

1 Introduction

The system of habitat classification presented here is designed for use in Phase 1 surveys of rural and urban areas throughout Great Britain.

This field manual has been written so that it can stand independently as a handbook for use during the fieldwork stage of a Phase 1 habitat survey. It also forms Part 2 of the complete *Handbook for Phase 1 habitat survey*. Part 1 contains information such as the rationale and history of Phase 1 survey, it gives advice on aspects such as planning the survey and it provides further guidance on habitat mapping and compiling target notes. The full handbook should be read before any fieldwork is undertaken.

The colour codes used in mapping are printed in the standard colours in the complete handbook, but this field manual is in black and white only. There are spaces in Appendix 1 which may be carefully coloured up in the appropriate manner, using the pattern provided in the full handbook and the Berol pencils detailed in Appendix 1, before this manual is taken into the field.

Habitats may be mapped using colour codes (Appendix 1), a hierarchical alphanumeric coding system, or lettered codes (Appendix 2). The use of codes for dominant species (Appendix 3) effectively provides a further subdivision of the habitat categories and is recommended for use wherever possible. Target notes give additional detail and they can also be used to clarify areas of difficulty in categorisation or mapping. Key words for use in compiling target notes are given in Appendix 4. A number of examples of target notes and a standard target note recording form are given in Appendices 5 and 6.

The full range of the classification should be used wherever possible, although this may at times prove difficult. For example, it may be impossible to determine the trophic status of a water body or to tell whether a fen is a basin, valley or flood-plain mire. For this reason, the recording of some habitat types is optional, as indicated in Appendix 1. In cases such as dry heath/acid grassland, the existence of vegetation mosaics has been recognised and a combination of colours is used in mapping.

Habitat definitions

This section provides definitions for each of the habitats which have been distinguished for the purpose of Phase 1 survey. The definitions given are based on those used by the NCC for surveying SSSIs (see Appendix 7). Appendix 8 shows the National Vegetation Classification (NVC) communities to be expected in each Phase 1 habitat category.

A Woodland and scrub

A1 Woodland

Woodland is defined as vegetation dominated by trees more than 5m high when mature, forming a distinct, although sometimes open, canopy. Dominant species should be coded and the understorey and ground layer target noted. Distinct blocks of woodland, whether broadleaved or coniferous, should be mapped separately wherever possible.

The definitions of the main categories are:-

broadleaved woodland: 10% or less conifer in the canopy;

coniferous woodland: 10% or less broadleaved in the canopy;

mixed woodland: 10-90% of either broadleaved or conifer in the canopy. The approximate proportions of the two types should be target noted.

If the cover of trees is less than 30% the area should be shown as scattered trees on the appropriate background colour. Where the cover is higher than 30% but there are sizeable open spaces or rides, these should be target noted to describe the ground flora.

Semi-natural woodland

Semi-natural woodland comprises all stands which do not obviously originate from planting. The distribution of species will generally reflect natural variations in the site and its soil. Both ancient and more recent stands are included. Woodland with both semi-natural and planted trees should be classified as semi-natural if the planted trees account for less than 30% of the canopy composition, but as plantation if more than 30% is planted. In cases where it is doubtful whether or not a wood should be classified as semi-natural, target notes giving details of origin and species composition are essential. For details of ancient woodland sites see Kirby *et al.* (1984).

2

The following should, amongst others, be included in the semi-natural category:-

> woods with planted standards in semi-natural coppice;
>
> mature plantations (more than about 120 years old) of native species growing on sites where those species are native and where there are semi-natural woodland ground flora and shrub communities;
>
> self-sown secondary stands of exotic species (for example sycamore, pine on southern heaths, holm oak on Isle of Wight);
>
> alder carr, and willow carr where the willows are more than 5m tall (although *Salix cinerea* should always be classified as scrub);
>
> well-established sweet-chestnut coppice (that is, over 25 years old);
>
> woods which have been completely underplanted, but where the planted trees do not yet contribute to the canopy;
>
> stands of young trees or coppice regrowth, even when less than 5 m.

Plantation woodland

All obviously planted woodland of any age should be included in this category, with the exception of those types mentioned previously. Orchards should be mapped by placing green hatching over the OS symbols (which should be added where missing), and target notes made giving tree species and details of any conservation interest. Ornamental tree gardens and arboreta should be included here, and target noted where necessary.

A2 Scrub

Scrub is seral or climax vegetation dominated by locally native shrubs, usually less than 5 m tall, occasionally with a few scattered trees. Dominant species should always be coded. The ground flora under scattered scrub should be coded or target noted.

The following should, amongst others, be included in this category:-

> *Ulex europaeus, Cytisus scoparius* and *Juniperus communis* scrub;
>
> stands of *Rubus fruticosus* and *Rosa canina*
>
> montane scrub with *Salix lapponum, S. lanata, S. myrsinites, S. arbuscula* or *S. phylicifolia;*

stands of mature *Crataegus monogyna*, *Prunus spinosa* or *Salix cinerea*, even if more than 5 m tall;

all willow carr less than 5 m tall; all *Salix cinerea* carr;

stands of *Myrica gale* more than 1.5 m tall.

The following should not be included in this category:-

very low *Salix herbacea* (see heathland, D), *Salix repens* (see dune slack, H6.4), or *Myrica gale* (see mire, E);

Ulex gallii or *Ulex minor* (see heathland D);

hedges (see J2);

stands of young trees or stump regrowth less than 5 m high, where these represent more than 50% of the immature canopy cover;

stands of introduced shrub species (see J1.4);

scrub on dunes (see H6.7).

A3 Parkland and scattered trees

Tree cover must be less than 30% to warrant inclusion in this category. For scattered trees over pasture (as in parkland), or over heath, bog, limestone pavement, etc, the green dot symbol should be superimposed on the appropriate habitat colour. The density of dots should be varied in proportion to the density of trees. Dominant species should be coded. Exotic trees should be target noted. Lines of trees forming windbreaks or avenues should be marked as a series of dots with the dominant species code.

A4 Recently-felled woodland

The only areas of felled trees which should be included in this category are those whose future land use is uncertain, for instance when it is not clear whether they are to be replanted or used for crops. The dominant species which have been felled should be coded and the codes placed in parentheses.

B Grassland and marsh

This category includes both areas of herbaceous vegetation dominated by grasses and certain wet communities dominated by *Juncus* species, *Carex* species, *Filipendula ulmaria* or by other marsh herbs. For grasslands where there is a greater than 25% cover of dwarf shrub heaths see heathland (D), for emergent stands of tall reed-grasses see swamp (F1), for coastal grasslands see saltmarsh (H2), dune (H6) and maritime cliff and slope (H8).

4

Most grasslands have been subjected to some degree of agricultural improvement by repeated grazing, mowing, fertilising, drainage or herbicide treatment. It is important to try to distinguish unimproved and semi-improved from improved grasslands. However, these grassland types form a continuum, so that it is not possible to define each with precision, especially as species critical for their definition are often only observable for a short season in the year. Agricultural improvement usually results in a decrease in the floristic diversity of the sward and dominance by a few quick-growing grasses such as *Lolium perenne*, *Holcus lanatus* and *Festuca rubra*. The resulting sward composition is likely to vary with intensity of treatment and with the composition of the original sward, so careful field training is necessary to define and maintain the boundaries between these categories. However, residual difficulties are bound to occur.

Grassy roadside verges, railway cuttings and embankments may be very important features, especially in intensively farmed areas. If they are wide enough they should be mapped as the appropriate grassland habitat. Narrow herb-rich verges should be shown by a broken orange line and target noted, if time permits. See also amenity grassland (J1.2).

Unimproved grassland

Unimproved grasslands are likely to be rare, especially in the lowlands. They may be rank and neglected, mown or grazed. They may have been treated with low levels of farmyard manure, but should not have had sufficient applications of fertiliser or herbicide, or have been so intensively grazed or drained, as to alter the sward composition significantly. Species diversity is often high, with species characteristic of the area and the soils and with a very low percentage of agricultural species.

In cases of doubt, map as semi-improved and target note the need for further information.

Semi-improved grassland

Semi-improved grassland is a transition category made up of grasslands which have been modified by artificial fertilisers, slurry, intensive grazing, herbicides or drainage, and consequently have a range of species which is less diverse and natural than unimproved grasslands. Such grasslands are still of some conservation value. Semi-improved grassland may originate from partial improvement of acid, neutral or calcareous grassland and should be mapped as such. However, it should be noted that improvement reduces the acid or calcareous character of the grassland, so that this is not always easy to distinguish in the field.

Species diversity will generally be lower than in unimproved grassland in the same area. If the signs of improvement listed under B4 are lacking, the grassland is likely to be semi-improved and should be mapped accordingly. Target notes should be made in all of the better quality sites. Surveyors should be aware of the species compositions indicative of semi-improved conditions in the locality of the survey. See also poor semi-improved grassland (B6).

B1 Acid grassland

Grassland in this category is often unenclosed, as on hill-grazing land, and occurs on a range of acid soils (pH less than 5.5). It is generally species-poor, and often grades into wet or dry dwarf shrub heath, although it must always have less than 25% dwarf shrub cover (see heathland, especially D5 and D6). Pioneer annual-rich calcifuge communities on dry sandy soils are included in this category, as are wet acidic grasslands typified by species such as *Juncus squarrosus* (but see marsh/marshy grassland, B5).

The following are indicative of acidic conditions when frequent or abundant: *Deschampsia flexuosa, Nardus stricta, Juncus squarrosus, Galium saxatile*, and *Rumex acetosella*.

B2 Neutral grassland

Typically enclosed and usually more intensively managed than acid or calcareous grassland (except on roadside verges), this category encompasses a wide range of communities occurring on neutral soils (pH 5.5-7.0).

The following are indicative of neutral conditions when frequent or abundant: *Alopecurus pratensis, Arrhenatherum elatius, Cynosurus cristatus, Dactylis glomerata, Deschampsia cespitosa, Festuca arundinacea* and *Festuca pratensis. Lolium perenne* may be present, but when abundant it is indicative of improved grassland (see B4).

Hay meadows will usually fall within this category. Surveyors should be aware that after cutting, a hay meadow can have the appearance of improved pasture as the new growth comes through.

Included in neutral grassland is a range of grasslands which are inundated periodically, permanently moist, or even water-logged (but see marsh/marshy grassland, B5). Examples are:-

inundated grassland with abundant *Glyceria* species, *Alopecurus geniculatus, Poa trivialis* and *Polygonum hydropiper;*

water meadows and alluvial meadows;

species-poor *Deschampsia cespitosa* grasslands and grazed *Juncus effusus/Juncus inflexus - Holcus lanatus/Deschampsia cespitosa* grasslands;

wet meadows or pastures where grasses are dominant in the sward (cf. marsh/marshy grassland, B5) but with species such as *Caltha palustris, Filipendula ulmaria, Valeriana* species, *Juncus* species or *Crepis paludosa* present.

B3 Calcareous grassland

These grasslands are often unenclosed, not managed intensively, and occur on calcareous soils (pH above 7.0). *Dryas octopetala* communities are included. Where the grass is tall, the dominant species is usually either *Brachypodium pinnatum* or *Bromus erectus*, whilst species indicative of short, close-grazed and species-rich calcareous turf are *Koeleria macrantha, Avenula pratensis, Sesleria albicans, Helianthemum nummularium, Sanguisorba minor* and *Thymus praecox*.

B4 Improved grassland

Improved grasslands are those meadows and pastures which have been so affected by heavy grazing, drainage, or the application of herbicides, inorganic fertilisers,

slurry or high doses or manure that they have lost many of the species which one could expect to find in an unimproved sward. They have only a very limited range of grasses and a few common forbs, mainly those demanding of nutrients and resistant to grazing. *Lolium perenne, Cynosurus cristatus, Trifolium repens, Rumex acetosa, Taraxacum officinale, Bellis perennis, Ranunculus acris* and *Ranunculus bulbosus* are typical of improved grassland, while stands of dock *Rumex* species, common nettle *Urtica dioica* and thistles *Cirsium* species indicate local enrichment of the soil by grazing animals.

The following signs usually indicate substantial improvement:-

bright green, lush and even sward, dominated by grasses (though poaching causes unevenness);

low diversity of forb species;

more than 50% *Lolium perenne, Trifolium repens* and other agricultural species.

Fields which have been reseeded in the past and have since become somewhat more diverse are included in this category, but recently reseeded monoculture grassland such as rye grass leys, with or without clover, should be classified under cultivated land (J1).

Most amenity grassland should also be classified under J1.

B5 Marsh/marshy grassland

This is a diffuse category covering certain *Molinia* grasslands, grasslands with a high proportion of *Juncus* species, *Carex* species or *Filipendula ulmaria*, and wet meadows and pastures supporting communities of species such as *Caltha palustris* or *Valeriana* species, where broadleaved herbs rather than grasses, predominate. The category differs from swamp (F1) in that the latter has a water table distinctly above the substratum for much of the year and is dominated by reed grasses or large sedges. Unlike marginal vegetation (F2), marsh/marshy grassland occurs on more or less level areas, rather than on the banks of watercourses. It differs from flush (E2) in that bryophytes are not a conspicuous component of the vegetation, also flushes always have a flow or seepage of water through them.

The following communities are included in marsh/marshy grassland:-

> vegetation with a greater than 25% cover of *Molinia caerulea*, on less than 0.5m of peat (cf. mire, E);

> vegetation with less than 25% dwarf shrub cover on peat less than 0.5 m deep (cf. heathland, D);

> vegetation with a greater than 25% cover of *Juncus acutiflorus, J. effusus, J. inflexus, Carex* species or *Filipendula ulmaria*, except for grazed *Juncus effusus - Holcus lanatus/Deschampsia cespitosa* grasslands, which should be classified under neutral grassland, B2;

> wet meadows and pastures where grasses are subordinate to forbs (cf. wet neutral grassland, B2). Such communities are often rich in plants such as *Caltha palustris, Filipendula ulmaria, Valeriana* species, *Crepis paludosa, Dactylorhiza* species, *Eupatorium cannabinum, Juncus* species and *Carex* species.

If *Sphagnum* is abundant, refer to the mire classification (E).

B6 Poor semi-improved grassland

Where there is a large amount of semi-improved grassland it may be useful to split this category into 'good semi-improved' and 'poor semi-improved', to facilitate re-survey of the better semi-improved grasslands at a later date. This sub-division is optional.

Good semi-improved grassland will have a reasonable diversity of herbaceous species, at least in parts of the sward, and is clearly recognisable as acid, calcareous or neutral in origin. Such grassland should be left in the semi-improved categories of acid, neutral and calcareous grassland (B1.2, 2.2 and 3.2). Poor semi-improved grassland will have a much more restricted list of species and, being more improved, it is more likely to resemble a species-poor neutral grassland, irrespective of its origin. This category (B6) should be marked SI and left uncoloured.

C Tall herb and fern

C1 Bracken

Areas dominated by *Pteridium aquilinum*, or with scattered patches of this species.

C2 Upland species-rich ledges

This ledge vegetation contains species such as *Angelica sylvestris, Filipendula ulmaria, Solidago virgaurea, Athyrium filix-femina, Trollius europaeus* and *Crepis paludosa*. Areas supporting this habitat are nearly always too small to map and consequently must be target noted.

C3 Other tall herb and fern

Tall ruderal (C3.1)

This category comprises stands of tall perennial or biennial dicotyledons, usually more than

25cm high, of species such as *Chamerion (Chamaenerion) angustifolium, Urtica dioica* and *Reynoutria japonica*. Dominant species should be coded. See also ephemeral/short perennial (J1).

Non-ruderal (C3.2)

Non-wooded stands of species such as *Oreopteris limbosperma, Athyrium felix-femina, Dryopteris* species or *Luzula sylvatica* should be included in this category. Dominant species should always be coded.

D Heathland

Heathland includes vegetation dominated by ericoids or dwarf gorse species, as well as 'heaths' dominated by lichens and bryophytes, dwarf forbs, *Carex bigelowii* or *Juncus trifidus*. Generally occurring on well-drained acid soils, heathland is further distinguished from mire (E) by being arbitrarily defined as occurring on peat less than 0.5m thick (but see flood-plain mire E3.3). Dominant species should always be coded. See also dune heath (H6.6) and coastal heathland (H8.5).

D1 Dry dwarf shrub heath

Vegetation with greater than 25% cover of ericoids or small gorse species in relatively dry situations forms this category. *Calluna vulgaris, Vaccinium myrtillus, Erica cinerea, Ulex minor* and *Ulex gallii* are typical of lowland dry dwarf shrub heath, whilst *Empetrum nigrum, Empetrum hermaphroditum, Arctostaphylos uva-ursi* and *Vaccinium vitis-idaea* are found in upland heaths. Acid heaths usually occur on deep podsols developed on base-deficient sands, gravels and clays. Basic heaths are much more restricted in extent, and may be recognised by the presence of herbs characteristic of chalk grassland and open habitats. See also wet dwarf shrub heath (D2), dry heath/acid grassland mosaic (D5) and dry modified bog (E1.4). Damp *Calluna* heath with *Sphagnum capillifolium* (mainly in western Scotland) should be included in this category and target noted.

D2 Wet dwarf shrub heath

As with dry dwarf shrub heath (D1), this vegetation type has more than 25% cover of ericoids and/or small *Ulex* species. However, it differs from D1 in that *Molinia caerulea* is often abundant and it generally contains some *Sphagnum compactum* or *Sphagnum tenellum* and less frequently other *Sphagna*. In transitions to mires, the proportion of *Sphagna* will increase and the species composition will change, often with *Sphagnum papillosum* and *Sphagnum subnitens* becoming more frequent. *Erica tetralix* is common in wet dwarf shrub heath and is often present in significant quantity. *Trichophorum cespitosum* is occasionally present at lower levels. Macrolichens such as

Cladonia portentosa (impexa), C. arbuscula and *C. uncialis* may be locally abundant. The abundance of *Molinia* and *Erica tetralix* decreases in the transition from wet to dry heath. See also wet heath/acid grassland mosaic (D6) and wet modified bog (E1.3).

D3 Lichen/bryophyte heath

This category comprises bryophyte and lichen-dominated heaths of mountain summits and lowland situations such as the East Anglian Breckland. Bryophytes and/or lichens must be dominant and there must be less than 30% vascular plant cover.

D4 Montane heath/dwarf herb

This is a rather diverse grouping of montane heath and snow-bed vegetation types. Included in this category are heaths dominated by *Carex bigelowii* and *Juncus trifidus*, also dwarf forb communities of *Alchemilla alpina, Silene acaulis, Sibbaldia procumbens* and *Saxifraga* species. Montane dwarf shrub heath should not be included, but should be classified under D1 or D2; *Dryas octopetala* communities should be classified under calcareous grassland (B3).

D5 Dry heath/acid grassland mosaic

This represents a common mixture of dry heath (D1) and acid

grassland (B1), to be found on hill and moorland, and the category has been specified only for ease of mapping. The relative proportions of each type of habitat should be target noted.

D6 Wet heath/acidic grassland mosaic

Vegetation mosaics similar to D5, but involving a mixture of wet heath (D2) with acid grassland (B1), make up this category. Again, the proportions of each habitat type should be target noted.

E Mire

Mires occur typically on deep peat (over 0.5 m thick) with the water table at or just below the surface, but flushes and springs on shallow or incipient peats are also included in this category.

The classification of peatlands has recently been revised (see NCC 1989) and the term bog is now restricted to ombrotrophic mires (blanket bog and raised bog), which are fed only by direct precipitation, unlike minerotrophic mires - fens (valley, flood-plain and basin mires), flushes and springs - which are fed by ground water or streams. The distinction between ombrotophic and minerotrophic mires is not always clear-cut and transitional examples should be target-noted. Furthermore, areas of minerotrophic mire may occur within blanket and raised mires; likewise, ombrotrophic areas may occur locally within fens. Examples of these should be target noted, but may be included within the major mire type for mapping purposes.

E1 Bog

Unmodified bog (blanket bog and raised bog) consists of *Sphagnum*-rich vegetation, lying on peat more than 0.5 m deep, with the water table at or just below the surface and no input of water from the surrounding land. Modified bog contains little or no *Sphagnum*.

Blanket bog (E1.6.1)

Blanket bog comprises *Sphagnum*-rich vegetation on deep peat, forming a blanket over both concave and convex surfaces, on level to moderately sloping ground in the uplands. It is widespread in the north and west of Britain, where it may be fragmentary or very extensive. The drainage is usually diffuse and undisturbed blanket bog often shows a hummock-and-hollow structure, with *Sphagnum*-rich pools in the hollows. Blanket bog includes watershed mires, saddle mires, terrace bog and valleyside mire and may also include other mire types, where these occur within a blanket bog complex.

This habitat category is used for relatively undamaged blanket bog, with *Sphagnum* usually abundant (typically *Sphagnum papillosum*, together with other species such as *Sphagnum magellanicum*). A wide range of ericoids, including *Calluna vulgaris, Erica tetralix, Vaccinium* species and *Empetrum* species, may be present, mainly on the hummocks, together with *Eriophorum vaginatum, Eriophorum angustifolium* and *Trichophorum cespitosum*. *Calluna* and/or *Eriophorum vaginatum* are often dominant over large areas, but various mixtures of species occur. Dominant species should be coded. Bog pool systems and areas of peat cutting, often characterised by the presence of *Sphagnum recurvum*, should be target noted or mapped as open water (G1.4) or bare peat (E4) if sufficiently large.

Significantly damaged blanket bog, in which *Sphagnum* is much reduced or absent, should be classified as modified bog (E1.7 or E1.8).

Raised bog (E1.6.2)

Raised bogs are found on estuarine flats, river flood plains and other level areas with impeded drainage in the lowlands, also at moderate altitudes, where they may grade into blanket mire. Many raised bogs overlie sites of glacial lakes which became infilled. In a classic raised bog, a structure now rare in Britain, the peat is several metres deep and has accumulated to form a distinctly raised dome, with peat depth greatest in the centre and decreasing towards the edges, which are marked by the more steeply sloping mire margin. Drainage tends to flow around the mire, forming a lagg stream, and the drier sloping margins of the mire may carry lagg woodland, which should be mapped as woodland.

Undamaged raised bog vegetation is very similar to that described under blanket bog (E1.6.1). Modification of raised bogs by draining, burning and peat-cutting can lead to the formation of wet modified bog and dry modified bog, which should be mapped as E1.7 or E1.8.

Wet modified bog (E1.7)

This category comprises modified bog vegetation with little or no *Sphagnum*, often with bare peat and patches of *Trichophorum cespitosum* and/or *Molinia caerulea*. Ericoids may be abundant, sparse or absent.

This vegetation is mainly found on drying and degraded blanket bogs and cut-over raised bogs. It may resemble wet heath (D2), but is distinguished by having a peat depth greater than 0.5 m. *Molinia*-dominated vegetation on deep peat is included in this category rather than in marshy grassland (B5).

Dry modified bog (E1.8)

The vegetation of dry modified bog is dominated by *Calluna vulgaris* and other ericoids, or by *Eriophorum vaginatum*, on peat more than 0.5m deep. *Sphagnum* is notably absent, but under the dwarf shrubs there may be a carpet of hypnoid mosses, with lichens such as *Cladonia portentosa* and *Cladonia arbuscula*. Where *Eriophorum vaginatum* is dominant, as on many Pennine blanket bogs, other species may be sparse or absent. Essentially dry heath vegetation (or cotton-grass moor) on deep peat, this habitat type is typical of areas of blanket bog or raised bog subjected to heavy grazing, burning and draining.

E2 Flush and spring

These types of minerotrophic mire are termed soligenous because they are associated with water movement. They may or may not form peat, but where they do, the peat is often less than 0.5 m deep. Flushes occur on gently-sloping ground, are often linear or triangular and may include small watercourses. They may be extensive or too small to map, in which case they should be target noted. Where flushes feed a fen (E3) they should be target noted and mapped as an integral part of the mire complex, unless they are very large and distinct, when they may be individually mapped.

Flushes typically have an open or closed ground layer of *Sphagnum* and/or other bryophytes, together with small sedges and *Juncus* species. The presence of a well developed bryophyte ground layer and the lack of dominant grasses distinguishes flush habitats from marshy grassland and from wet acid, neutral and calcareous grasslands. Thus, a habitat with *Juncus effusus* over herbs and grasses is a marsh/marshy grassland (B5). Complex mosaics of grassland and flush are quite common, particularly in the uplands, and should be mapped according to the most prevalent habitat, with the proportions of each recorded in a target note.

Flushes may be acid, neutral (mesotrophic) or basic. These categories are not always easy to distinguish. In cases of doubt use the magenta colour code only and target note.

Acid/neutral flush (E2.1)

These typically support species-poor vegetation consisting of a *Sphagnum* carpet overlain by *Carex* or *Juncus* species. Characteristic moss species include *Sphagnum recurvum*, *S. palustre* and *S. auriculatum*. Overlying vegetation may consist of small *Carex* species (*Carex echinata*, *C. nigra* or *C. curta*), *Carex rostrata*, *Juncus acutiflorus*, *J. effusus*, *J. squarrosus*, or *Eriophorum angustifolium*. Dominant species should be coded.

Basic flush (E2.2)

Basic flushes typically support a carpet of pleurocarpous brown mosses, often without *Sphagnum*, overlain by a conspicuous small sedge layer, *Carex flacca*, *Schoenus nigricans* or a mixed-herb layer. Characteristic pleurocarpous mosses include *Scorpidium*, *Campylium*, *Drepanocladus* and *Calliergon* species, whilst characteristic herbs include *Eleocharis quinqueflora*, *Eriophorum latifolium* and *Carex lepidocarpa*.

Bryophyte-dominated spring (E2.3)

This habitat occurs only in the immediate vicinity of up-wellings and it usually consists of spongy mats or small mounds dominated by bryophytes such as *Cratoneuron* or *Philonotis* species. Areas which fall within this category are normally too small to map and should be target noted. Flushes occurring downslope of a spring should be mapped if they are large enough.

E3 Fen

Fens are defined as minerotrophic mires, usually over peat more than 0.5 m deep (but see E3.3). The water table is at or just below the surface. Three main types of fen can be distinguished, using topographical rather than vegetational criteria. These are valley mire, which, because there is obvious water flow, is classified as soligenous, and basin and flood- plain mires, which have impeded drainage and are termed

topogenous. However, the distinction between these three mire types is not always clear in the field, so for Phase 1 mapping purposes their identification is optional.

'Poor fen' contains acid water (pH 5 or less) and short vegetation with a high proportion of *Sphagnum*. 'Rich fen' contains more calcareous water (pH above 5), *Sphagnum* is often absent and the vegetation usually includes patches of tall plants and species such as *Juncus subnodulosus*, *Schoenus nigricans* and *Carex lepidocarpa*, characteristic of base-rich situations. Where acid or basic fen can be identified, this should be made clear in a target note and basic fen should be indicated by the code 'B'.

Where there are very wet areas containing tall swamp vegetation such as *Phragmites australis* or large sedges, these should be target noted as swamp (F1), or marked as patches of sky blue, if large enough to map within the area delineated as fen. Parts of the mire dominated by marsh (fen meadow) or carr should be mapped or target noted as grassland (B5), woodland (A1) or scrub (A2). Springs and small flushes which feed or lie within a fen, should be treated as an integral part of the mire system and target noted (see E2). Areas of bog within a fen and patches of degraded fen should also be target noted.

Valley mire (E3.1)

A valley mire develops along the lower slopes and floor of a small valley and receives water from springs and seepages on the valley sides, feeding a central watercourse. Such a fen can be distinguished from a flush because the former is a complex, whereas a flush is a discrete single feature, usually of limited extent.

Valley mires are often dominated by acidophilous vegetation containing *Sphagnum* species, *Carex* species and ericoids. However, vegetation typical of base-rich conditions can also occur, for instance *Schoenus nigricans* and *Juncus subnodulosus*. Floating mats of mosses and sedges may be present. Acid watercourses often contain *Hypericum elodes* and *Potamogeton polygonifolius*.

Basin mire (E3.2)

This type of fen develops in a waterlogged basin and contains very little open water. The water table within the basin is level, but small flushes may occur around the edges and there is a limited through-flow of water.

The vegetation may be dominated by *Sphagnum* species, together with *Carex rostrata* and ericoids, or by tall swamp plants such as *Phragmites australis*, *Schoenoplectus (Scirpus) lacustris*, *Typha* species and, in base-rich situations, *Cladium mariscus*.

Flood-plain mire (E3.3)

This type of fen forms on a river or stream flood-plain which is waterlogged and, typically, inundated periodically. The substrate may be peat, mineral or a mixture of both. The range of vegetation types is similar to that of a basin mire (E3.2).

E4 Bare peat

Patches of bare peat more than 0.25 ha in extent (that is, approximately 50 m x 50 m) should be mapped. Peat hagging and areas of eroding peat haggs should be target noted. Commercial peat-workings are included in this category.

F Swamp, marginal and inundation

This habitat category is defined as emergent or frequently inundated vegetation, occurring over peat or mineral soils. The depth of water at the time of survey, or seasonal variation in water level, if known, should be target noted, also the nature of the substrate. Note that this category differs from mire (E) and from marsh/marshy grassland (B5) in having the water table distinctly above the level of the substrate for most of the year.

F1 Swamp

Swamp contains tall emergent vegetation typical of the transition between open water and exposed land. Swamps are generally in standing water for a large part of the year, but may occasionally be found on substrates that are seldom immersed, as in the later stages of the seral succession to marshy grassland.

Species composition varies according to the trophic status of the water, the substrate type, etc. Note that vegetation dominated by *Molinia caerulea*, *Filipendula ulmaria*, mosses, small *Carex* species or *Juncus* species, should be classified as marsh/marshy grassland (B5) or flush (E2), as appropriate. Swamp vegetation includes both mixed and single-species stands of *Typha* species, *Phragmites australis*, *Phalaris arundinacea*, *Glyceria maxima*, *Carex paniculata*, *C. acutiformis*, *C. rostrata* or other tall sedge. Single-species stands are usually found in deeper water and should be indicated with species codes.

Strips of swamp vegetation narrower than 5m bordering watercourses should be classified as marginal vegetation (F2.1).

F2 Marginal and inundation

Marginal vegetation (F2.1)

This category encompasses all narrow strips of emergent vegetation occurring on the (often steep) margins of lowland watercourses, where the water table is permanently high. Bands of tall vegetation wider than 5 m should be classified as swamp (F1). Marginal vegetation is typically open and contains plants such as *Glyceria* species, *Rorippa* species, *Apium nodiflorum*, *Berula erecta*, *Oenanthe* species, *Galium palustre*, *Nasturtium officinale*, *Myosotis* species, *Veronica* species, *Alisma* species, *Sparganium erectum*, *Carex riparia*, *Juncus effusus* and *Juncus inflexus*, also small stands of taller plants such as *Phragmites australis*, *Typha* species and *Phalaris arundinacea*. Areas of such vegetation will be too small to map, so should be target noted.

Inundation vegetation (F2.2)

This category includes open and innately unstable communities that are subject to periodic inundation, as found on sorted or unsorted silts, sands and gravels of river beds and islands and on the draw-down zone around pools, lakes and reservoirs. A wide variety of species occur in such communities, including *Polygonum* species, *Juncus bulbosus*, *Bidens* species, *Agrostis stolonifera* and *Alopecurus geniculatus*, as well as many ruderal species.

G Open water

Open water is defined as water lying beyond the limits of swamp or emergent vegetation, although it may contain submerged, free-floating or floating-leaved vegetation. The dominant species of any such vegetation should be coded, and the salinity of the water, whether fresh or brackish, indicated if possible. Where aquatic vegetation is present in quantity but there is insufficient room to code all abundant species, a target note should be provided. For those wishing to provide details of the trophic status of the water, Table 1 gives the characteristics of each type (see also Palmer 1989).

G1 Standing water

Standing water includes lakes, reservoirs, pools, flooded gravel pits, ponds, water-filled ditches, canals and brackish lagoons.

G2 Running water

Running water comprises rivers and streams. The direction of flow should be indicated by an arrow. If survey is needed at a more detailed level than for Phase 1, refer to *Surveys of wildlife in river corridors* (NCC 1985). This draft methodology includes a comprehensive classification of bank and open water habitats, a recording card and instructions on the preparation of habitat maps.

Table 1 Classification of standing and running waters

	Physical characteristics	Typical plant species
1 Eutrophic	Water often strongly discoloured by algae. pH usually over 7. Substrate often highly organic mud.	*Lemna* spp. *Myriophyllum spicatum* *Potamogeton pectinatus* *Ceratophyllum* spp. *Zannichellia palustris* *Ranunculus circinatus* *Polygonum amphibium* *Chara* spp. *Nuphar lutea* *Ranunculus penicillatus* var. *calcareus* is typical of flowing waters
2 Mesotrophic	Water sometimes discoloured by planktonic algae. pH usually around or slightly below neutral	*Potamogeton gramineus* *P. obtusifolius* *P. perfoliatus* *Callitriche hermaphroditica* *Nitella* spp. *Nuphar lutea* *Nymphaea alba*
3 Oligotrophic	Water very clear, plankton sparse. pH usually less than 7. Substrate rocky, sandy or peaty.	*Potamogeton polygonifolius* *Myriophyllum alterniflorum* *Juncus bulbosus* *Scirpus fluitans* *Subularia aquatica* *Lobelia dortmanna* *Isoetes lacustris* *Sparganium angustifolium* *Callitriche hamulata.* Flowing waters dominated by bryophytes.

4 Dystrophic	Water usually peat-stained. pH very low (3.5 - 5.5). Alkalinity very low (up to 2mg/l CaCO₃)	*Sphagnum* spp. *Juncus bulbosus* *Potamogeton polygonifolius* Macrophyte flora very restricted.
5 Marl/tufa	May be eutrophic, mesotrophic or (very rarely) oligotrophic. Water very clear. Alkalinity at least 100mg/l CaCo₃. Powdery yellow-brown deposit of marl covers substrate in lakes. Highly calcareous streams deposit tufa.	*Chara* spp. *Myriophyllum spicatum* *Potamogeton lucens*
6 Brackish	Most brackish systems are coastal, but a few are inland, with salinity derived from artificial sources such as mine drainage, or from residues of ancient marine incursions in peaty areas. Conductivity 1,250 to 50,000 µmhos.	Flora very restricted. Slightly saline waters - *Potamogeton pectinatus* *P. pusillus* *Myriophyllum spicatum* *Zannichellia palustris* *Ceratophyllum submersum* *Ranunculus baudotii* *Enteromorpha* spp. More saline waters - *Ruppia* spp, fucoids, *Zostera* spp.

H Coastland

Coastal lagoons should be classified as standing water (G1.6).

H1 Intertidal

The codes for *Zostera*, green algal beds or brown algal beds should, where appropriate, be superimposed over the relevant Ordnance Survey symbols (mud /sand; shingle /cobbles; boulders/rocks)

H2 Saltmarsh

Saltmarsh/dune interface (H2.3)

Vegetation peculiar to this area, characterised by species such as *Frankenia laevis* or *Suaeda fruticosa*, should be mapped wherever large enough, and always target noted.

Scattered plants (H2.4)

The dominant species should be coded.

Dense/continuous (H2.6)

Dominant species should be coded, particularly noting *Spartina* where it is abundant. Areas of inland saltmarsh should be included in this category.

H3 Shingle/gravel above high-tide mark

Target note any vascular plants or lichen vegetation that may occur.

H4 Boulders/rocks above high-tide mark

Target note as for H3.

H5 Strandline vegetation

This type of vegetation occurs as an open community on the drift line and is characterised by species such as *Cakile maritima*, *Honkenya peploides*, *Rumex crispus*, *Salsola kali*, *Atriplex* species and *Beta vulgaris* ssp. *maritima*. In contrast to fore dunes, *Elymus farctus* (*Agropyron junceiforme*) is characteristically sparse or absent. Target note where feasible, stating whether the substrate is shingle or rock.

H6 Sand dune

Dune slack (H6.4)

Dune slacks are valleys or hollows between dune ridges, where the water table is close to the surface for at least several months in the year, leading to marshy vegetation. *Ammophila arenaria* is usually absent. Characteristic species are *Salix repens*, *Hydrocotyle vulgaris*, *Dactylorhiza* species and *Epipactis palustris*. Saline slacks should be classified as saltmarsh (H2).

Dune grassland (H6.5)

All grassland occurring on consolidated and flattened dunes should be classified in this category. Generally, little

Ammophila arenaria will be present. Machair should be included here.

Dune heath (H6.6)

All heathland occurring on consolidated and flattened dunes should be included in this category. *Calluna* is usually the dominant ericoid, with *Erica cinerea* and *Erica tetralix* also common. *Carex arenaria* is often present and lichens, particularly *Cladonia* species, are often abundant. Occasionally, juniper may be present. Use yellow crosses for scattered heath.

Dune scrub (H6.7)

All scrub occurring on consolidated and flattened dunes should be classified in this category. *Hippophae rhamnoides* is a characteristic species. Use green crosses for scattered scrub.

Open dune (H6.8)

This category comprises the three early successional stages of dune formation, less stable and with lower vegetation cover than H6.4-H6.7.

Fore dune: unstable, usually low ridges of sand on the foreshore, often with a very open plant cover. *Elymus farctus* is strongly characteristic, often dominant, and sometimes the only species present; *Honkenya peploides, Atriplex* species and *Cakile maritima* are typical associated species; *Ammophila arenaria* may be present in small quantities, but should not be dominant.

Yellow dune: partially stabilized ridges of sand lying between fore and grey dunes, with a marked but incomplete plant cover, nearly always dominated by *Ammophila arenaria*, although *Leymus (Elymus) arenarius* and/or *Elymus farctus* may be common; a variety of small herbs may be present.

Grey dune: stable ridges of sand, almost completely vegetated. The vegetation is very variable in species composition; *Ammophila arenaria* is usually present, but not dominant, mosses and lichens may be frequent. Grey dune can be distinguished from fixed dune by being markedly hilly or undulating, and by the sand not being fully consolidated.

H8 Maritime cliff and slope

Maritime hard cliff (H8.1)

These are cliffs formed of rock (including chalk) with less than 10% vascular plant cover. The type of rock should be target noted.

Vegetated cliffs should be mapped using the relevant vegetation code and target noted.

Maritime soft cliff (H8.2)

These are cliffs formed of mud or clay with less than 10% vascular plant cover. The type of substrate should be target noted.

Crevice and ledge vegetation (H8.3)

This category comprises vegetation, occasionally sparse, but covering at least 10% of the cliff surface, occurring in crevices or on ledges on steep cliffs. The communities present should be described with a target note, taking care to record whether the vegetation is influenced by the use of the cliffs by birds, as may be indicated by species such as *Beta vulgaris*. Vegetation occurring in the splash zone at the base of cliffs should be included here.

Coastal grassland (H8.4)

These are grasslands which include maritime species and which occur on shallow slopes or level areas by the sea, often on cliff tops (but see dune grassland - H6.5). Indicator species include *Scilla verna*, *Plantago maritima* and *Armeria maritima*. *Festuca rubra* is often dominant. Other species may include *Hieracium pilosella*, *Anthyllis vulneraria*, *Lotus*

corniculatus, *Galium verum* and *Thymus praecox*.

Coastal heathland (H8.5)

All heathlands which include maritime species and which occur on shallow slopes, or even level areas, by the sea should be included in this category (but see dune heath - H6.6). Indicator species include *Scilla verna*, *Armeria maritima*, *Jasione montana*, *Plantago maritima* and *Plantago coronopus*. *Calluna vulgaris* is often dominant; *Erica cinerea* and dwarf *Ulex* species are frequently present. Coastal heathland often occurs just inland of coastal grassland, and like that category, frequently occurs at the top of cliffs.

I   Rock exposure and waste

This grouping includes both natural and artificial exposed rock surfaces where these are almost entirely lacking in vegetation, as well as various forms of excavations and waste tips. Significant communities of mosses, lichens and ferns growing on walls or rock ledges should be target noted. See also maritime cliff and slope (H8).

I1 Natural exposures

Inland cliff (I1.1)

This category is defined as rock surfaces over 2 m high and sloping at more than 60°. Vegetated cliffs with more than 10% vascular plant cover are not included, but should be mapped using the relevant vegetation code, and target noted as necessary.

Scree (I1.2)

Scree is defined as an accumulation, usually at the foot of a cliff, of weathered rock fragments of all sizes, mostly angular in shape. This category includes large boulders (boulder scree) which should be mapped using enlarged red dots.

Limestone pavement (I1.3)

This comprises a near horizontal surface, usually of Carboniferous Limestone, which is irregularly corrugated and furrowed by solution and often cut by deeper and more regular fissures (grikes), which correspond to naturally occurring joints within the rock.

Other exposure (I1.4)

Exposed rock on mountain tops and in river beds should, for example, be included in this category.

Cave (I1.5)

Any natural recess, large enough to enter and with a complete ceiling, should be mapped as cave and any features of interest target noted. Large crevices and deep narrow gullies should not be included here, but should be mapped under 'other'.

I2 Artificial exposures and waste tips

The boundaries of quarries, spoil heaps, mines or refuse tips should be outlined in red. Covering vegetation, if abundant, should be coded as appropriate, under grassland, scrub, etc, or target noted if sparse.

Quarry (I2.1)

Excavations such as gravel, sand or chalk pits and stone quarries should be included in this category. Target note the mineral or ore which has been, or is being, extracted. If the site is water-filled, map as open water and target note previous use.

Spoil (I2.2)

Includes abandoned industrial areas and tips of waste material such as coal mine spoil and slag. Spoil heaps within quarries should be included in I2.1. Target note the type of spoil.

Mine (I2.3)

Mark the area on the map and target note any features of interest.

Refuse-tip (I2.4)

Target note any vegetation of interest, if it covers an area too small to map, and code the dominant species.

J Miscellaneous

Features such as parks, gardens, golf courses and railway cuttings or embankments are not listed as separate habitat types, since they are clearly marked as such on Ordnance Survey maps. The colour codes for the appropriate habitat type (for example grassland, woodland or scrub) should, however, be superimposed over the feature on the Ordnance Survey map.

J1 Cultivated/disturbed land

Arable (J1.1)

This includes arable cropland, horticultural land (for example nurseries, vegetable plots, flower beds), freshly-ploughed land and recently reseeded grassland, such as rye grass and rye-clover leys, often managed for silage.

Amenity grassland (J1.2)

This comprises intensively managed and regularly mown grasslands, typical of lawns, playing fields, golf course fairways and many urban 'savannah' parks, in which *Lolium perenne*, with or without *Trifolium repens*, often predominates. The sward composition will depend on the original seed mixture used and on the age of the community. Herbs such as *Bellis perennis*, *Plantago major* and *Taraxacum officinale*

may be present. If the amenity grassland has a sward rich in herbs, it may be possible to classify it as semi-improved acidic, neutral or calcareous grassland, as appropriate. In such cases, the area concerned should be mapped as the specific grassland type and its amenity use target noted.

Ephemeral/short perennial (J1.3)

Short, patchy plant associations typical of derelict urban sites, quarries and railway ballast, should be classified here. The land must be freely draining, and usually has shallow stony soil. The vegetation typically lacks a clear dominant species, but consists of a mixture of low-growing plants, often less than 25 cm high, such as *Plantago major*, *Ranunculus repens*, *Trifolium repens*, *Medicago lupulina*, *Tussilago farfara*, *Leucanthemum vulgare* and *Senecio* species, or of taller species such as *Sisymbrium* or *Melilotus* species. Parts of fields containing similar communities, such as areas around gates, should not be included, but should be classified as grassland (B). See also tall ruderal (C3.1).

Introduced shrub (J1.4)

This is vegetation dominated by shrub species that are not locally native, whether planted or self-sown. Common introduced shrubs include species of *Buxus*, *Cornus*, *Laurus*, *Ligustrum*, *Rhododendron* and *Symphoricarpus*. Formal beds of shrubs such as of *Hypericum calycinum, Cotoneaster,* heaths and dwarf conifers should be included here. Introduced shrubs forming an understorey in woodland should be mapped as woodland (A1) and target noted. Introduced shrub on sand dunes should be classified as dune scrub (H6.7). See also scrub (A2).

J2 Boundaries

Although a key to field boundaries is supplied, time constraints often preclude the mapping of boundaries. Nevertheless, the conservation value of hedges should not be overlooked and it is recommended that at least the better examples should be mapped and target noted, particularly in lowland areas. Species-rich hedges should be differentiated from species-poor ones by the use of the zig-zag symbol. Fences are usually of little significance, as their wildlife value is low, but recording new boundaries and removed boundaries may be important. A clear decision should be made as to the types of boundary to be mapped and consistency should be maintained.

Guidance on recording grassy road verges, railway cuttings and embankments is given under the grassland section (B). Where they are dominated by trees or scrub they should be categorised as woodland (A) and mapped if very broad, otherwise simply target noted.

Intact hedge (J2.1)

Intact hedges are entire and more-or-less stock-proof.

Defunct hedge (J2.2)

Hedges in which there are gaps and which are no longer stock-proof fall into this category.

Hedgerow with trees (J2.3)

The frequency of cross-hatching should be varied to indicate the density of trees. Windbreaks should be classified under A3.

Species-rich hedges

These have a diversity of native woody species and a good hedgerow bottom flora.

Wall (J2.5)

Significant communities of mosses, lichens or ferns growing on walls may be target noted, particularly in built-up areas.

Ditch (J2.6)

Only ditches which appear to be dry for most of the year should be included in this category. Wet ditches are mapped as standing water (G1) or possibly swamp (F1).

Boundary removed (J2.7)

Use spaced crosses on the appropriate Ordnance Survey symbol.

Earth bank (J2.8)

The ditch/bank systems found on ancient woodland sites may be included here, as should sea walls constructed of natural materials.

J3 Built-up areas

Caravan site (J3.4)

Hatching may be used as an overlay on the appropriate semi-natural habitat colour, for instance where the site is on coastal grassland or in woodland.

Sea wall (J3.5)

Only sea walls constructed of artificial materials (for example concrete) should be included here. Others should be mapped as earth banks (J2.8).

Buildings (J3.6)

Map unmarked new buildings or built-up areas and colour those already shown on the Ordnance Survey maps. Agricultural, industrial and domestic buildings should all be coloured in solid black. There is no need to distinguish between them.

J4 Bare ground

Any type of bare soil or other substrate should be included here where not already covered (compare bare peat E4, intertidal H1, shingle H3, boulders and rocks H4, Dunes H6, maritime cliff H8 and natural rock exposure I). Target note extensive or otherwise important areas of bare ground.

J5 Other habitat

Draw a black line around any habitat not encompassed by the classification and describe it in a target note.

Phase 1 survey habitat classification, hierarchical alphanumeric reference codes and mapping colour codes

A | Woodland and scrub

1 Woodland

   1 Broad-leaved

     1 Semi-natural   Green

     2 Plantation   Green

   2 Coniferous

     1 Semi-natural   True green

     2 Plantation   True green

   3 Mixed

     1 Semi-natural   Green over true green

     2 Plantation   Green and true green

2 Scrub

   1 Dense/continuous   Green

   2 Scattered   Green

3 Parkland/scattered trees

   *1 Broad-leaved   Green

   *2 Coniferous   True green

   *3 Mixed   Green and true green

4 Recently-felled woodland

   *1 Broad-leaved   Green

   *2 Coniferous   True green

   *3 Mixed   Green and true green

| B | Grassland and marsh |

1 Acid grassland
1 Unimproved   [   ]   Orange

2 Semi-improved   [SI]   Orange

2 Neutral grassland
1 Unimproved   [   ]   Orange

2 Semi-improved   [SI]   Orange

3 Calcareous grassland
1 Unimproved   [   ]   Orange

2 Semi-improved   [SI]   Orange

4 Improved grassland   [I]   No colour

5 Marsh/marshy grassland   [   ]   Purple over orange

*6 Poor semi-improved grassland (optional)   [SI]   No colour

C | Tall herb and fern

1 **Bracken**	1 Continuous		Terra cotta
	2 Scattered		Terra cotta
2 **Upland species-rich ledges**			Terra cotta target note
3 **Other**	1 Tall ruderal		Terra cotta
	2 Non-ruderal		Terra cotta

D | Heathland

1 **Dry dwarf shrub heath**	1 Acid		Yellow ochre
	2 Basic	B	Yellow ochre
2 **Wet dwarf shrub heath**			Purple over yellow ochre
3 **Lichen/bryophyte heath**			Yellow ochre
4 **Montane heath/dwarf herb**			Yellow ochre
5 **Dry heath/acid grassland mosaic**			Orange over yellow ochre
6 **Wet heath/acid grassland mosaic**			Purple and orange over yellow ochre

| E | Mire |

1 Bog

*6 Sphagnum bog

   *1 Blanket bog   [   ]   Purple

   *2 Raised bog   [RB]   Purple

  *7 Wet modified bog   [   ]   Purple

  *8 Dry modified bog   [   ]   Purple

2 Flush and spring

  1 Acid/neutral flush   [   ]   Magenta

  2 Basic flush   [B]   Magenta

  3 Bryophyte-dominated spring   [   ]   Target note

*3 Fen

   [   ]   Magenta over purple

  Optional codings:

  Basic   [B]   Magenta over purple

  *1 Valley mire   [VM]   Magenta over purple

  *2 Basin mire   [BM]   Magenta over purple

  *3 Flood-plain mire   [FPM]   Magenta over purple

*4 Bare peat   [   ]   Purple

Swamp, marginal and inundation

1 Swamp			Sky blue
2 Marginal and inundation	1 Marginal vegetation		Sky blue + target note
	2 Inundation vegetation		Sky blue

Open water

1 Standing water — Indigo blue

Optional codings:

1 Eutrophic	E	Indigo blue
2 Mesotrophic	M	Indigo blue
3 Oligotrophic	O	Indigo blue
4 Dystrophic	D	Indigo blue
5 Marl	C	Indigo blue
6 Brackish (*includes saline lagoons)	B	Indigo blue

2 Running water — Indigo blue

Optional codings:
1 Eutrophic   E
2 Mesotrophic   M
3 Oligotrophic   O
*4 Dystrophic   D
*5 Marl/tufa   C
*6 Brackish   B

H	Coastland

1 Intertidal

    1 Mud/sand
    2 Shingle/cobbles } Ordnance Survey symbols
    3 Boulders/rocks

    Codings for intertidal:
    *1 Zostera beds   Zo
    *2 Green algal beds   Ga
    *3 Brown algal beds   Ba

2 Saltmarsh

    3 Saltmarsh/dune interface   [ ]   Pink + target note

    4 Scattered plants   [ ]   Pink

    *6 Dense continuous   [ ]   Pink

3 Shingle above high tide mark } Ordnance Survey symbols

4 Boulders/rocks above high tide mark

5 Strandline vegetation   Target note

6 Sand dune

    4 Dune slack   [ ]   Indigo blue over flesh

    5 Dune grassland   [ ]   Orange over flesh

    6 Dune heath   [ ]   Yellow ochre over flesh

    7 Dune scrub   [ ]   Green over flesh

    *8 Open dune   [ ]   Flesh

8 Maritime cliff and slope

    1 Hard cliff   [ ]   Scarlet red

    2 Soft cliff   [ ]   Scarlet red

    *3 Crevice/ledge vegetation   Target note

    *4 Coastal grassland   [C]   Orange

    *5 Coastal heathland   [C]   Yellow ochre

I	Rock exposure and waste

1 Natural

1 Inland cliff

   1 Acid/neutral   Scarlet red

   2 Basic   Scarlet red

2 Scree

   1 Acid/neutral   Scarlet red

   2 Basic   Scarlet red

3 Limestone pavement   Scarlet red

4 Other exposure

   1 Acid/neutral   Scarlet red

   2 Basic   Scarlet red

5 Cave   Scarlet red

2 Artificial

1 Quarry   Scarlet red

2 Spoil   Scarlet red

3 Mine   Scarlet red

*4 Refuse-tip   Scarlet red

J | Miscellaneous

1 Cultivated/ disturbed land

    *1 Arable   A   No colour

    *2 Amenity   A   Canary
       grassland   yellow

    *3 Ephemeral/short   Black
       perennial

    *4 Introduced shrub   Terra cotta

2 Boundaries
(mapping optional)

    1 Intact hedge

      *1 Native species-   Green
         rich

      *2 Species-poor   Green

    2 Defunct hedge

      *1 Native species-   Green
         rich

      *2 Species-poor   Green

    3 Hedge and trees

      *1 Native species-   Green
         rich

      *2 Species-poor   Green

    '4 Fence   Black

    5 Wall   Scarlet red

    6 Dry ditch   Indigo blue

    *7 Boundary removed   Black

    *8 Earth bank   Black

3 **Built-up areas**	4 Caravan site		Black
	*5 Sea wall (artificial material)		Black
	*6 Buildings		Black
4 **Bare ground**			Black
5 **Other habitat**			Black + target note

her mapping aids

Target note	⊙	Red 0.35 mm Rotring pen
Dividing line between habitats where no boundary is marked on the map	– – –	Black 0.35 mm Rotring pen
Phase 1 survey boundary (delimits area of current field survey)	– – –	Red 0.5 mm Rotring pen
Habitat information not obtainable because of restricted access	NA	

tes

Use only the standard colours in the Berol Verithin series, obtainable from stationers or from Berol Ltd., Oldmeadow Road, King's Lynn, Norfolk, PE30 4JR.

VT 01 Black	VT 46 Orange	VT 05 Indigo blue
VT 49 Pink	VT 08 Sky blue	VT 51 Purple
VT 25 Flesh	VT 55 Scarlet red	VT 31 Green
VT 66 Terra cotta	VT 32 True green	VT 89 Yellow ochre
VT 45 Magenta	VT 80 Canary yellow	

Shading should be even and fairly light so as not to obscure underlying detail on the printed map, but not so light as to cause difficulty in distinguishing or reproducing the colours. Hatching should be evenly spaced and consistent in direction.

Code the dominant species wherever possible, using codes in Appendix 3.

*indicates where Phase 1 alphanumeric reference codes differ from those used in either the 1984 NCC/RSNC classification or the 1982 SSSI mapping system (Appendix 7).

See Appendix 2 for a comparison of alphanumeric and mnemonic lettered habitat codes.

Appendix 2

Habitat codes for use on monochrome field maps and fair maps

				Alphanumeric code	Lettered code
A	**Woodland and scrub**				
1	Woodland				
	Broadleaved	-	semi-natural	A1.1.1	BW
		-	plantation	A1.1.2	PBW
	Coniferous	-	semi-natural	A1.2.1	CW
		-	plantation	A1.2.2	PCW
	Mixed	-	semi-natural	A1.3.1	MW
		-	plantation	A1.3.2	PMW
2	Scrub	-	dense/continuous	A2.1	DS
		-	scattered	A2.2	SS
3	Parkland/	-	broad-leaved	A3.1	SBW
	scattered	-	coniferous	A3.2	SCW
	trees	-	mixed	A3.3	SMW
4	Recently-	-	broad-leaved	A4.1	FB
	felled	-	coniferous	A4.2	FC
	woodland	-	mixed	A4.3	FM
B	**Grassland and marsh**				
1	Acid	-	unimproved	B1.1	AG
	grassland	-	semi-improved	B1.2	SAG
2	Neutral	-	unimproved	B2.1	NG
	grassland	-	semi-improved	B2.2	SNG
3	Calcareous	-	unimproved	B3.1	CG
	grassland	-	semi-improved	B3.2	SCG
4	Improved grassland			B4	I
5	Marsh/marshy grassland			B5	MG
6	Poor semi-improved			B6	SI

Tall herb and fern

1	Bracken	- continuous	C1.1	CB
		- scattered	C1.2	SB
2	Upland species-rich ledges		C2	Target note
3	Other	- tall ruderal	C3.1	TR
		- non-ruderal	C3.2	NR

Heathland

1	Dry dwarf	- acid	D1.1	ADH
	shrub heath	- basic	D1.2	BDH
2	Wet dwarf shrub heath		D2	WH
3	Lichen/bryophyte heath		D3	LH
4	Montane heath/dwarf herb		D4	MH
5	Dry heath/acid grassland mosaic		D5	DGM
6	Wet heath/acid grassland mosaic		D6	WGM

Mire

1	Bog	- blanket bog	E1.6.1	BB
		- raised bog	E1.6.2	RB
		- wet modified	E1.7	WB
		- dry modified	E1.8	DB
2	Flush/Spring	- acid/neutral	E2.1	AF
		- basic	E2.2	BF
		- bryophyte dom.	E2.3	Target note
3	Fen	- valley mire	E3.1	VM)
		- basin mire	E3.2	BM) B where
		- flood-plain	E3.3	FPM) basic
4	Bare peat		E4	P

Swamp, marginal and inundation

1	Swamp		F1	SP
2	Marginal/	- marginal	F2.1	MV
	inundation	- inundation	F2.2	IV

G Open water

1	Standing water	-	eutrophic	G1.1	SWE
		-	mesotrophic	G1.2	SWM
		-	oligotrophic	G1.3	SWO
		-	dystrophic	G1.4	SWD
		-	marl	G1.5	SWC
		-	brackish	G1.6	SWB
2	Running water	-	eutrophic	G2.1	RWE
		-	mesotrophic	G2.2	RWM
		-	oligotrophic	G2.3	RWO
		-	dystrophic	G2.4	RWD
		-	marl	G2.5	RWC
		-	brackish	G2.6	RWB

H Coastland

1	Intertidal	-	mud/sand	H1.1	O.S. symbol
		-	shingle/cobbles	H1.2	O.S. symbol
		-	boulders/rocks	H1.3	O.S. symbol
			Zostera beds	H1.(1-2).1	Zo
			green algal beds	H1.(1-3).2	Ga
			brown algal beds	H1.(1-3).3	Ba
2	Saltmarsh	-	saltmarsh/dune interface	H2.3	Target note
		-	scattered plants	H2.4	SSM
		-	dense/continuous	H2.6	DSM
3	Shingle above high tide mark			H3	O.S. symbol
4	Boulders/rocks above high tide mark			H4	O.S. symbol
5	Strandline vegetation			H5	Target note
6	Sand-dune	-	dune slack	H6.4	DW
		-	dune grassland	H6.5	DG
		-	dune heath	H6.6	DH
		-	dune scrub	H6.7	DX
		-	open dune	H6.8	OD
8	Maritime cliff and slope	-	hard cliff	H8.1	HC
		-	soft cliff	H8.2	SC
		-	crevice/ledge vegetation	H8.3	Target note
		-	coastal grassland	H8.4	SG+target n●
		-	coastal heathland	H8.5	SH+target n●

I Rock exposure and waste

1 Natural

Inland cliff	-	acid/neutral	I1.1.1	AC
	-	basic	I1.1.2	BC
Scree	-	acid/neutral	I1.2.1	AS
	-	basic	I1.2.2	BS
Limestone pavement			I1.3	LP
Other exposure	-	acid/neutral	I1.4.1	AR
	-	basic	I1.4.2	BR
Cave			I1.5	CA

2 Artificial

	-	quarry	I2.1	Q
	-	spoil	I2.2	S
	-	mine	I2.3	MI
	-	refuse-tip	I2.4	R

J Miscellaneous

1 Cultivated/disturbed land

	-	arable	J1.1	A
	-	amenity grassland	J1.2	AM
	-	ephemeral/short perennial	J1.3	ESP
	-	introduced shrub	J1.4	IS

2 Boundaries

Hedges	- intact	- species-rich	J2.1.1	RH
		- species-poor	J2.1.2	PH
	- defunct	- species-rich	J2.2.1	RH-
		- species-poor	J2.2.2	PH-
	- with trees	- species-rich	J2.3.1	RHT
		- species-poor	J2.3.2	PHT
Fence			J2.4	F
Wall			J2.5	W
Dry ditch			J2.6	DD
Boundary removed			J2.7	X
Earth bank			J2.8	EB

3 Built-up areas

	-	caravan site	J3.4	CS
	-	sea wall	J3.5	SWALL
	-	buildings	J3.6	Shade black

4 Bare ground

		J4	BG

5 Other habitat

		J5	Target note

Note
The lettered habitat codes listed in this appendix differ in some respects from the letters overlaid on the colour mapping codes (see Appendix 1).

Appendix 3

Dominant species codes

Ac	*Acer campestre*
Ap	*Acer pseudoplatanus*
Ah	*Aesculus hippocastanum*
Aeu	*Agrimonia eupatoria*
At	*Agrostis capillaris (A.tenuis)*
Agc	*Agrostis curtisii (A.setacea)*
Agt	*Agrostis stolonifera*
Aip	*Aira praecox*
Ajr	*Ajuga reptans*
Aa	*Alchemilla alpina*
Alu	*Allium ursinum*
Ag	*Alnus glutinosa*
Alg	*Alopecurus geniculatus*
Ama	*Ammophila arenaria*
An	*Anemone nemorosa*
As	*Angelica sylvestris*
Ao	*Anthoxanthum odoratum*
Asy	*Anthriscus sylvestris*
Auu	*Arctostaphylos uva-ursi*
Ae	*Arrhenatherum elatius*
Av	*Artemisia vulgaris*
Ast	*Aster tripolium*
Aff	*Athyrium filix-femina*
Apr	*Avenula pratensis (Helictotrichon pratense)*
Apb	*Avenula pubescens (Helictotrichon pubescens)*
Bet	*Betula* sp(p)
Bpe	*Betula pendula*
Bpu	*Betula pubescens*
Bsp	*Blechnum spicant*
Bp	*Brachypodium pinnatum*
Bs	*Brachypodium sylvaticum*
Bm	*Briza media*
Be	*Bromus erectus (Zerna erecta)*
Bxs	*Buxus sempervirens*
Ce	*Calamagrostis epigejos*
Cac	*Calamagrostis canescens*
Cv	*Calluna vulgaris*
Cap	*Caltha palustris*
Cx	*Carex* sp(p)
Cxaa	*Carex acuta*
Cxac	*Carex acutiformis*
Cxar	*Carex arenaria*
Cxb	*Carex bigelowii*
Cxe	*Carex elata*
Cxf	*Carex flacca*
Cxl	*Carex laevigata*
Cxn	*Carex nigra*
Cxo	*Carex otrubae*
Cxpa	*Carex paniculata*
Cxrm	*Carex remota*
Cxri	*Carex riparia*
Cxro	*Carex rostrata*
Cxv	*Carex vesicaria*
Cb	*Carpinus betulus*
Cs	*Castanea sativa*
Cn	*Centaurea nigra*
Cha	*Chamerion (Chamaenerion) angustifolium*
Cop	*Chrysosplenium oppositifolium*
Cl	*Circaea lutetiana*
Cirs	*Cirsium* sp(p)
Car	*Cirsium arvense*
Ch	*Cirsium helenioides (C. heterophyllum)*
Cp	*Cirsium palustre*
Civ	*Cirsium vulgare*
Clm	*Cladium mariscus*
Com	*Conium maculatum*
Cva	*Clematis vitalba*
Cos	*Cornus sanguinea (Thelycrania sanguinea)*
Ca	*Corylus avellana*
Cot	*Cotoneaster* sp(p)
Cm	*Crataegus monogyna*
Crc	*Crepis capillaris*
Cyc	*Cynosurus cristatus*
Cys	*Cytisus scoparius (Sarothamnus scoparius)*
Dg	*Dactylis glomerata*
Dd	*Danthonia decumbens*
Dc	*Deschampsia cespitosa*
Df	*Deschampsia flexuosa*
Dp	*Digitalis purpurea*
Do	*Dryas octopetala*
Dr	*Dryopteris* sp(p)
Ddl	*Dryopteris dilatata*
El	*Elymus* sp(p)
Ef	*Elymus farctus (Agropyron junceiforme)*
En	*Empetrum nigrum*
Ep	*Epilobium* sp(p)
Eq	*Equisetum* sp(p)
Egf	*Equisetum fluviatile*
Eqs	*Equisetum sylvaticum*
Ec	*Erica cinerea*
Et	*Erica tetralix*
Erio	*Eriophorum* sp(p)

Era	*Eriophorum angustifolium*	Liv	*Ligustrum vulgare*
Ev	*Eriophorum vaginatum*	Lvu	*Limonium vulgare*
Ee	*Euonymus europaeus*	Lnv	*Linaria vulgaris*
Fs	*Fagus sylvatica*	Lp	*Lolium perenne*
Fo	*Festuca ovina*	Lpc	*Lonicera periclymenum*
Fp	*Festuca pratensis*	Lzp	*Luzula pilosa*
Fr	*Festuca rubra*	Ls	*Luzula sylvatica*
Fu	*Filipendula ulmaria*	Lmm	*Lysimachia nummularia*
Fv	*Fragaria vesca*	Md	*Malus domestica*
Fa	*Frangula alnus*	Maq	*Mentha aquatica*
Fe	*Fraxinus excelsior*	Mp	*Mercurialis perennis*
Glt	*Galeobdolon luteum (Lamiastrum galeobdolon)*	Mc	*Molinia caerulea*
		Mym	*Mycelis muralis*
Gap	*Galium aparine*	Mg	*Myrica gale*
Gsx	*Galium saxatile*	Ns	*Nardus stricta*
Gp	*Geranium pratense*	Nos	*Narthecium ossifragum*
Gro	*Geranium robertianum*	Noth	*Nothofagus* sp.
Gs	*Geranium sylvaticum*	NuN	*Nuphar/Nymphaea* sp(p)
Gu	*Geum urbanum*	Nl	*Nuphar lutea*
Gh	*Glechoma hederacea*	Na	*Nymphaea alba*
Gf	*Glyceria* sp(p) (except *maxima)*	Odv	*Odontites verna*
		Ol	*Oreopteris limbosperma*
Gm	*Glyceria maxima*	Oxa	*Oxalis acetosella*
Hp	*Halimione portulacoides*	Pas	*Pastinaca sativa*
Hh	*Hedera helix*	Pet	*Petasites hybridus*
Hc	*Helianthemum chamaecistus (H. nummularium)*	Pha	*Phalaris arundinacea*
		Phl	*Phleum pratense* agg.
Hsp	*Heracleum sphondylium*	Pc	*Phragmites australis (P. communis)*
Hr	*Hippophae rhamnoides*		
Hx	*Holcus* sp(p)	Px	*Picea* sp(p)
Hl	*Holcus lanatus*	Pia	*Picea abies*
Hn	*Hyacinthoides non-scripta (Endymion non-scriptus)*	Pin	*Pinus* sp(p)
		Psyl	*Pinus sylvestris*
Hyp	*Hypericum perforatum*	Pl	*Plantago lanceolata*
Ia	*Ilex aquifolium*	Pm	*Plantago major*
Ig	*Impatiens glandulifera*	Po	*Poa* sp(p)
Ip	*Iris pseudacorus*	Pbis	*Polygonum bistorta*
Ju	*Juncus* sp(p)	Pop	*Populus* sp(p)
Ja	*Juncus acutiflorus*	Pot	*Populus tremula*
Jar	*Juncus articulatus*	Pom	*Potamogeton* sp(p)
Je	*Juncus effusus*	Pans	*Potentilla anserina*
Ji	*Juncus inflexus*	Ppal	*Potentilla palustris*
Jm	*Juncus maritimus*	Psn	*Poterium sanguisorba (Sanguisorba minor)*
Jsq	*Juncus squarrosus*		
Js	*Juncus subnodulosus*	Pv	*Prunella vulgaris*
Jt	*Juncus trifidus*	Pa	*Prunus avium*
Jc	*Juniperus communis*	Pap	*Prunus avium/padus*
Lx	*Larix* sp(p)	Pnc	*Prunus cerasus* agg.
Lxd	*Larix decidua*	Pdn	*Prunus domestica*
Le	*Lemna* sp(p)	Pp	*Prunus padus*
Lv	*Leucanthemum vulgare (Chrysanthemum leucanthemum)*	Ps	*Prunus spinosa*
		Pgm	*Pseudotsug. menziesii*
		Pt	*Pteridium aquilinum*

Puc	*Puccinellia* sp(p)	Sr	*Sedum rosea*
Py	*Pyrus* sp(p)	Sj	*Senecio jacobaea*
Qu	*Quercus* sp(p)	Sea	*Sesleria albicans (Sesleria*
Qp	*Quercus petraea*		*caerulea)*
Qr	*Quercus robur*	Sia	*Silene acaulis*
Rs	*Ranunculus* sp(p)	Sdi	*Silene dioica*
Rfm	*Ranunculus flammula*	Sd	*Solanum dulcamara*
Rr	*Ranunculus repens*	Soa	*Sonchus arvensis*
Rll	*Reseda luteola*	Sa	*Sorbus aria*
Rj	*Reynoutria japonica*	Sac	*Sorbus aucuparia*
Rl	*Racomitrium lanuginosum*	Se	*Sparganium erectum*
Rc	*Rhamnus catharticus*	Sp	*Spartina* sp(p)
Rhin	*Rhinanthus minor* agg.	Sph	*Sphagnum* sp(p)
Rp	*Rhododendron ponticum*	Sme	*Stellaria media*
Ros	*Rosa* sp(p)	Sf	*Suaeda fruticosa*
Rch	*Rubus chamaemorus*	Sup	*Succisa pratensis*
Rf	*Rubus fruticosus* agg.	Tb	*Taxus baccata*
Ri	*Rubus idaeus*	Tsn	*Teucrium scorodonia*
Rx	*Rumex* sp(p)	Til	*Tilia* sp(p)
Ra	*Rumex acetosella*	Tic	*Tilia cordata*
Sal	*Salicornia* sp(p)	Tip	*Tilia platyphyllos*
Sx	*Salix* sp(p)	Tiv	*Tilia vulgaris (Tilia europaea)*
Sxa	*Salix alba*	Tc	*Trichophorum cespitosum*
Sxar	*Salix arbuscula*		*(Scirpus cespitosus)*
Sxau	*Salix aurita*	Tr	*Trifolium repens*
Sxc	*Salix caprea*	Tff	*Tussilago farfara*
Sxci	*Salix cinerea*	Ty	*Typha* sp(p)
Sxf	*Salix fragilis*	Ta	*Typha angustifolia*
Sxh	*Salix herbacea*	Tl	*Typha latifolia*
Sxl	*Salix lapponum*	Ul	*Ulex* sp(p)
Sxm	*Salix myrsinites*	Ue	*Ulex europaeus*
Sxp	*Salix pentandra*	Ug	*Ulex gallii*
Sxpu	*Salix purpurea*	Umi	*Ulex minor*
Sxr	*Salix repens*	Um	*Ulmus* sp(p)
Sxt	*Salix triandra*	Umg	*Ulmus glabra*
Sxv	*Salix viminalis*	Ump	*Ulmus procera*
Sn	*Sambucus nigra*	Ud	*Urtica dioica*
Sang	*Sanguisorba officinalis*	Vm	*Vaccinium myrtillus*
Sne	*Sanicula europaea*	Vv	*Vaccinium vitis-idaea*
Saa	*Saxifraga aizoides*	Vd	*Valeriana dioica*
Slt	*Schoenoplectus (Scirpus)*	Vth	*Verbascum thapsus*
	lacustris	Vl	*Viburnum lantana*
Sl	ssp *lacustris*	Vop	*Viburnum opulus*
St	ssp *tabernaemontani*	Vh	*Viola hirta*
Sc	*Schoenus nigricans*	Vip	*Viola palustris*
Sv	*Scilla verna*	Zo	*Zostera* sp(p)
Sm	*Scirpus maritimus*	Zoa	*Zostera angustifolia*
Sno	*Scrophularia nodosa*	Zon	*Zostera noltii*

Group abbreviations

Ba	-	Brown algae (*Fucus* etc in coastal waters)
Brys	-	Bryophytes (other than *Sphagnum*)
Cons	-	Conifers
Exos	-	Exotic shrubs
Ga	-	Green algae (including *Ulva, Enteromorpha* and filamentous algae, in saline and fresh waters)
Licn	-	Lichens

Notes

1 These codes should be used for mapping dominant species, by superimposing the species code on the appropriate colour code for the habitat (see Appendix 1), where there is sufficient room on the map. The occurrence of dominant species on small areas and non-dominant species of interest should be recorded in target notes.

2 New codes may be devised for any species not given on this list, as long as care is taken to ensure that they do not duplicate existing symbols and the NCC is consulted. New codes should be sent to the England Headquarters, Nature Conservancy Council, Northminster House, Peterborough, PE1 1UA, so that they can be incorporated into the centrally-held master list.

Appendix 4

Key words and status categories for target notes

Key words

General	confidential further Phase 1 survey required Phase 2 equired photograph taken wildlife corridor		pest control restoration scrub control tree felling unenclosed
Geology/ topography	acid geology chalk clay eroded gravel landslide limestone sand sandstone silt steep slope soil stability subsidence substratum water table	Recreation	**boating** caravans climbing fishing footpath horse riding hunting off-road vehicles public pressure shooting skiing walking watersports
Management (general)	active cultivated burnt cut derelict disused drained fenced failed flooded enclosed game management stock grazed irrigated mown neglected	**Damage (general)**	afforestation bracken invasion construction deer/rabbit damage destruction disturbance drainage dumping erosion extraction fertiliser fire flood overgrazing herbicide liming peat cutting

	pesticide (other than herbicide)	fragmentary vegetation
	pollution	fungus
	scrub invasion	insect
	sewage	introduced species
	silage leachate	invasive species
	slurry	invertebrate
	tree invasion	lichen
	undergrazing	mammal
		NVC community
		orchid
Flora/	algae	reptile
fauna	amphibia	seabird
	bat	species-poor
	bird	species-rich
	bryophyte	succession
	butterfly	wader
	dragonfly	wildfowl
	fern	
	fish	

A   Woodland and
    Scrub

ancient
boundary bank
canopy
carr
clearing
coppice
dead wood
even age
fallen trees
firebreak
forestry
ground flora
high forest
mixed age
no regeneration
orchard
planted
pollard
recent
regeneration
rides
scrubby
shrub layer
standard
underplanted

B   Grassland and
    Marsh

alluvial
ant hills
hay meadow
ley
litter layer
mole hills
permanent
rabbit grazed
ridge and furrow
short
tall
time of cutting
water meadow

C   Tall herb and fern

dry
mixed
monospecific
wet

D   Heathland

fire break
grouse moor
rabbit grazed
snow bed

E   Mire

drying
hagg
hollow
hummock
lagg
patterned
pool

F   Swamp, marginal
    and inundation

drying
fluctuating water
level
bare mud
seasonal
watertable

G   Open Water

abstraction
artificial bank
canal
canalised
dam
deep
ditch
drawdown
dredged
fast flow
fish farming
fishery
floating vegetation
gravel pit

hydrosere
island
lake/loch
marina
natural bank
ornamental
phytoplankton bloom
pond
reservoir
riffle
rocky
shallow
slow flow
submerged vegetatio
tarn
turbid
waterfall
weed cutting

H Coastland	I Rock exposure and waste	J Miscellaneous
accreting	bare (of vegetation)	boundary removed
bait-digging	coal	cemetery
coastal defence	fossil	churchyard
eroding	gully	construction
machair	geological interest	dry ditch
reclamation	hard	earth bank
rock pool	mineral	fallow
sea wall	ore	field margin
seepage	removal	garden
spray zone	soft	hedge
stable	vegetated ledge	industry
		ley
		pipeline
		reseeded
		tunnel
		verge
		wall

Status categories

Sites

Ancient monument
AONB (AGLV in Scotland)
Bird sanctuary
Commonland
Country Park
County Trust reserve (including RSNC holdings)
Crown Estate
Forest Nature Reserve
GCR (Geological Conservation Review) site
Heritage Coast
ISR (Invertebrate Site Register) site
LNR (Local Nature Reserve)
MNR (Marine Nature Reserve)
Ministry of Defence
National Scenic Area (Scotland)
National Park
National Trust
NCR (Nature Conservation Review) site
NNR (National Nature Reserve)
Private nature reserve
Public Open Space
Ramsar site
RSPB reserve
SPA (EC Special Protection Area)
SSSI
Wildfowl refuge
Wildlife Scheme
Woodland Trust reserve

Species

alien
breeding
characteristic
endemic
indicator
rarity – county
        – international
        – local
        – national (Red Data Boc
        – nationally scarce
        – notable
        – regional
        – WCA (Wildlife &
          Countryside Act
          Schedules 1, 5 or 8)

Appendix 5

Hypothetical examples of target notes

OS Sheet: SK 52 SW

1x1 km grid square: 5122

510227	15.6.83 JF, RT	Neutral grassland; unimproved; mown. *Alopecurus pratensis* dominant. Frequent *Orchis morio* and *Primula veris.* Small wet areas with *Caltha palustris.* Owner at adjacent farmhouse.
	7.80	Hay meadow card in Scientific File SK52.
513224	13.5.83 JF, PK	Scroggs Wood. Broadleaved, semi-natural wood, coppiced; probably ancient. Dominated by *Quercus robur.* Shrub layer poor: *Corylus, Sorbus aucuparia.* Ground flora poor: *Holcus lanatus* and *Deschampsia cespitosa* dominant. *Dryopteris dilatata* and *Trientalis europaea* frequent. Apparently neglected for about 40 years; unenclosed; stock grazed; no regeneration.
513227	18.5.83 LW, RT	Blue Tarn. Shallow, oligotrophic standing water with valley mire adjacent. Emergent vegetation dominated by *Juncus* spp and *Menyanthes trifoliata.* Grades into *Sphagnum* + *Eriophorum* bog in the south, and into *Salix* carr in the north. Good for dragonflies, including *Anax imperator* and *Erythromma najas.* Phase 2 survey required; site may warrant special protection.

| 4 | 515229 | 20.7.83
LW, RT | Fragmentary marginal vegetation, occurring along 100 m of the north bank of the river Ugie.
Dominant species: *Glyceria plicata* and *Juncus effusus*.
Breeding mallard, moorhen and reed bunting. |
| 5 | 517220 | 8.82 | Home Farm pond.
Great crested newt - County Wildlife Trust survey, 1982. |

Appendix 6 Standard recording forms
Map sheet record

Survey title			
Map sheet No.*		10 km sq.	Orig. ref. no.
Surveyor(s)		Survey date(s)	
County (counties)			
L.A. district(s)			
Area (ha) surveyed			
Total area (ha) surveyed		Location of records	

Notes: (include e.g. use of aerial photos, supplementary sources of information, level of detail, habitats and areas omitted)

Habitat measurements Habitat name	or Standard Phase 1 alphanumeric code	Area (ha)	Length (m)

* 1:10,000 or 1:25,000 sheet no. or, where measurements are on 1km square basis, substitute grid ref.

53

Target note record

Survey title		Site name	
Grid ref.	10 km sq	Orig. ref. no.	
Surveyor(s)		Survey date	
Conservation status (if any)		SSSI code	
Country	L.A. district	Vice county no.	

Habitat(s) included	Name	or alphanumeric code
dominant		
other		

Target note (mention supplementary source(s) of information, if used)

Key words by habitat (include management, damage, etc.)

A Woodland	B Grassland	C Tall herb/fern	D Heathland	E Mire

F Swamp etc.	G Open water	H Coastland	I Rock etc.	J Miscellaneous

Species recorded		
Species name (preferably Latin)	Standard species code	Notes (include habitat code, species status, abundance*)
Dominant plant species		
Other species of interest (nationally scarce, notable, indicator, etc.)		

* Express plant abundance (where not dominant) as abundant/frequent/occasional/rare

Appendix 7

The NCC/RSNC habitat classification (revised 1984)

First level hierarchy	Second level hierarchy	Third level hierarchy	Fourth level hierarchy
A Woodland and scrub	1 Woodland	1 Broadleaved 2 Coniferous 3 Mixed	1 Semi-natural 2 Plantation
	2 Scrub	1 Dense/ continuous 2 Scattered	[1 Acidic] [2 Neutral] [3 Basic]
	3 Parkland and scattered trees 4 Recently felled woodland		
B Grassland	1 Acidic 2 Neutral 3 Basic/ calcareous	1 Unimproved 2 Semi-improved	1 Upland 2 Lowland
	4 Improved/ reseeded 5 Marshy grassland	1 Upland 2 Lowland	
C Tall herb and fern	1 Bracken	1 Continuous 2 Scattered	
	2 Upland spp. - rich vegetation		
	3 Other tall herb or fern	1 Ruderal/ ephemeral 2 Other	
D Heathland	1 Dry dwarf shrub heath	1 Acidic 2 Basic	1 Upland 2 Lowland
	2 Wet dwarf shrub heath 3 Lichen/	1 Upland 2 Lowland	

				bryophyte heath		
			4	Montane heath/dwarf herb		
			5	Dry heath/ acidic grass mosaic		
			6	Wet heath/ acidic grass mosaic		
E	Bog and flush	1	Bog	1	Blanket bog	1	Open *Sphagnum* carpets
				2	Upland raised bog	2	*Eriophorum vag.* and other bog veg. over *Sphagnum*
				3	Lowland raised bog	3	Mosaic of 1 and 2
				4	Valley bog	4	Bog veg. over *Sphagnum* (no *Eriophorum vag.*)
				5	Basin mire	5	Mosaic of 1 and 4
						6	Wet heath over deep peat (no *Sphagnum*)
						7	Dry heath over deep peat (no *Sphagnum*)
						8	Bare peat
						9	Open bog pools
		2	Flush	1	Acidic flush		
				2	Basic flush		
				3	Bryophyte-dominated flush		
F	Swamp and fen/ inundation communities	1	Swamp and fen	1	Single sp. dominant swamp		
				2	Tall fen vegetation		
		2	Open marginal/ inundation communities	1	Fragmentary marginal vegetation		
				2	Inundation communities	–	

57

G	Open water	1	Standing water	1	Eutrophic	[1	Small ponds]		
				2	Mesotrophic	[2	Ponds, etc <0.5 ha]		
				3	Oligotrophic	[3	Lakes 0.5 ha]		
				4	Dystrophic	[4	Large lakes > 5 ha]		
				5	Marl	[5	Canals and ditches]		
				6	Brackish	[6	Reservoirs]		
		2	Running water	1	Eutrophic	[1	Spring and small stream <1 m wide]		
				2	Mesotrophic	[2	Streams and rivers 1-3 m wide]		
				3	O;ogotrophic	[3	Rivers > 3 m wide]		
				4	Marl	(5)*			
				5	Brackish	(6)*			
H	Coastland	1	Intertidal	1	Mud/sand	1	*Zostera* beds	(4)*	
				2	Shingle/cobbles	2	Green algal beds	(6)*	
				3	Boulders/rocks	3	Brown algal beds	(6)*	
		2	Saltmarsh	1	*Spartina*				
				2	Other sp.(p)				
				3	Saltmarsh/ dune interface				
				4	Scattered plants				
				5	Inland saltmarsh				
		3	Shingle						
		4	Rock/ boulders						
		5	Strandline vegetation						
		6	Sand dune	1	Fore dune				
				2	Yellow dune				
				3	Grey dune				
				4	Dune slack				
				5	Dune grassland				
				6	Dune heath				
				7	Dune scrub				
		7	Lagoon	-					
		8	Maritime cliff	1	Hard	1	Crevice/ledge vegetation	(3)*	
				2	Soft	2	Seacliff		

						grassland	(4)*
					3	Seacliff heath	(5)*
					4	Bird cliff vegetation	(6)*

I	Rock	1	Natural rock exposures and caves	1	Inland cliff	1	Acidic
				2	Scree	2	Basic
				3	Limestone pavement		
				4	Other	1	Acidic
				5	Cave	2	Basic
				[6	Mountain top]		
				[7	Riverine]		
				[8	Ravine]		
		2	Artificial rock exposures	1	Quarry		
				2	Spoil heap		
				3	Mine		
J	Other	1	Cultivated land				
		2	Boundary	1	Intact hedge		
				2	Defunct hedge		
				3	Hedgerow with trees		
				4	Fence		
				5	Wall		
				6	[Dry] Ditch		
				(7)**			
		3	Building	1	Agricultural (incl. forestry)		
				2	Industrial		
				3	Domestic		
				4	Caravans		
		4	Bare ground				
		5	Others				

K   Marine

Note

This classification is very similar to the 1982 NCC SSSI habitat mapping system, (which follows the 1982 version of the NCC/RSNC classification). The differences are indicated on the table thus:-

[ ]   =   Not included in the SSSI habitat mapping system.

*   =   Numbered differently in the SSSI habitat mapping system - the SSSI number codes in parentheses.

J(7)** =   Boundary removed - an additional category in the SSSI habitat mapping system.

Appendix 8.

Relationship between Phase 1 habitat categories and National Vegetation Classification Communities in *British plant communities* Volumes 1-5.

This table is not definitive, but gives provisional guidance only. Few of the correspondences are exact and many NVC communities correspond to more than one Phase 1 category. Some correspondences are only at sub-community level (details not given here). A fuller version will be made available via the JNCC website in due course.

code	Phase 1 category	Principal associated NVC communities (not necessarily confined to Phase 1 category)
A	**Woodland and scrub**	
A1	Woodland	
A1.1.1	Broadleaved woodland - semi-natural	W4-12, W14-17; W2 & W19 (birch - dominated)
A1.1.2	Broadleaved woodland - plantation	W8-12, W14-17
A1.2.1	Coniferous woodland - semi-natural	W13, W18; W16 (with self-sown pine)
A1.2.2	Coniferous woodland - plantation	W6, W10-11, W16, W18
A1.3.1	Mixed woodland - semi-natural	W8, W8-18
A1.3.2	Mixed woodland - plantation	W8-11, W14-18
A2	Scrub	W1-7, W19-25; M15 & M25 (with tall *Myrica* dominant)
A3	Parkland and scattered trees	Various grassland, heathland, scrub and woodland types
B	**Grassland and marsh**	
B1.1	Acid grassland - unimproved	U1-6; SD10-11 (inland stands)
B1.2	Acid grassland - semi-improved	U4 and others
B2.1	Neutral grassland - unimproved	MG1-5, MG8-10, MG12
B2.2	Neutral grassland - semi-improved	MG1, MG6, MG9-10, MG12-13
B3.1	Calcareous grassland - unimproved	CG1-14, OV37
B3.2	Calcareous grassland - semi-improved	CG2-4
B4	Improved grassland	MG6-7
B5	Marsh/marshy grassland	MG8, MG10, M22-28
B6	Poor semi-improved grassland	MG6 and others
C Tall herb and fern		
C1	Bracken	U20, W25
C2	Upland species-rich ledges	U17
C3.1	Other tall herb and fern - tall ruderal	OV24-27
C3.2	Other tall herb and fern - non ruderal	U16, U18-19, U21, OV38
D Heathland		
D1.1	Dry dwarf shrub heath - acid	H1-4, H6, H8-10, H12-22
D1.2	Dry dwarf shrub heath - basic	H6-8, H10
D2	Wet dwarf shrub heath	M15-16, H5
D3	Lichen/bryophyte heath	H1, H13-14, H17, H19-20, U1, U10, SD11
D4	Montane heath/dwarf herb	U7-15
D5	Dry heath/acid grassland	H1-4, H8-10, H12, H15-18, U1-6
D6	Wet heath/acid grassland	M15-16, U2-6

E	Mire	
E1.6.1	Blanket bog	M1-3, M15, M17-20
E1.6.2	Raised bog	M1-3, M17-20
E1.7	Wet modified bog	M1-3, M15, M20, M25
E1.8	Dry modified bog	M3, M19-20, H9-10, H12
E2.1	Flush and spring - acid/neutral flush	M4, M6-7, M14, M21, M29
E.2.2	Flush and spring - basic flush	M10-14, M34-36
E2.3	Flush and spring - bryophyte-dominated spring	M31-33, M37-38
E3.1	Fen - valley mire	M21, M29 and various other Mire and Swamp types
E3.2	Fen - basin mire	M1-9, S24-28 and various other Mire and Swamp types
E3.3	Fen - flood plain mire	Various Mire and Swamp types

F	Swamp, marginal and inundation	
F1	Swamp	S1-22, S27-28
F2.1	Marginal/inundation - marginal	S1-23, S26-28, A1-3, A10, A16, A19-21
F2.2	Marginal/inundation - inundation	A2, A10, A16, A19-21, M30, MG11, MG13, OV28-36, S23

G	Open water	
G1.1	Standing water - eutrophic	A1-8, A10-12, A15-16, A19-20
G1.2	Standing water - mesotrophic	A2-3, A7-11, A13-16, A19-20
G1.3	Standing water - oligotrophic	A7, A9-10, A14, A16, A22-24
G1.4	Standing water - dystrophic	A7, A9, A24
G1.5	Standing water - marl	A5, A8, A11, A15-17, A19-20
G1.6	Standing water - brackish	A1, A3, A6, A12, A21, SM1-2
G2.1	Running water - eutrophic	A1-2, A5, A11-12, A15-17, A19-20
G2.2	Running water - mesotrophic	A2, A8-9, A11, A13-20
G2.3	Running water - oligotrophic	A7-9, A14, A18
G2.4	Running water - dystrophic	A7, A9
G2.5	Running water - marl	A5, A11, A15-17, A19-20
G2.6	Running water - brackish	A1, A21, SM2

H	Coastland	
H1.1	Intertidal - mud/sand	SM1-3
H2	Saltmarsh	SM4-28, MG11-12
H3	Shingle/gravel above high-tide mark	SD1-4, MC6 and various maritime grassland, heath and scrub types
H4	Rocks/boulders above high-tide mark	MC1-7
H5	Strandline vegetation	SD1-3, MC6
H6.4	Dune slack	SD13-17 and others
H6.5	Dune grassland	SD7-12, SD19, U1 and various other grassland types
H6.6	Dune heath	H1, H10, H11, M15-16
H6.7	Dune scrub	SD18, W1
H6.8	Open dune	SD4-7, SD10-11, SD19
H8.3	Crevice/ledge vegetation	MC1-8, SM16
H8.4	Coastal grassland	MC6-12
H8.5	Coastal heathland	H7-8

I	**Rock exposure and waste**	
I1.1.1	Inland cliff - acid/neutral	U18, U21 and various others
I1.1.2	Inland cliff - basic	OV39-41
I1.2.1	Scree - acid/neutral	U18, U21
I1.2.2	Scree - basic	OV38, OV40
I1.3	Limestone pavement	Fragments of CG9, M10, M26-27, MG OV38-40, W9 and others

J	**Miscellaneous**	
J1.1	Cultivated/disturbed land - arable	OV1-19
J1.2	Cultivated/disturbed land - amenity grassland	OV23
J1.3	Cultivated/disturbed land - ephemeral/short perennial	OV18-22
J2.5	Wall	OV39, OV41-42

Appendix 9.

Technical developments and other changes since 1990 - key points.

Section 2.2 and 3.2-3.8

There have been major developments in the area of electronic data handling and storage, notably the use of Geographical Information Systems (GIS), and in remote sensing techniques. Hand-held GPS (Global Positioning System) is now a great boon for surveyors especially in more difficult terrain, and equipment allowing electronic mapping and data entry in the field combined with GPS is now becoming available. Further developments in these areas will undoubtedly revolutionise fieldwork in the coming years.

Section 6.3 (also 2.9.1)

The standard taxonomic reference for vascular plants is now Stace (1997) and the companion field flora (Stace 1999).

Section 9.3 (also 1.2) and Appendix 8

The National Vegetation Classification (NVC) has now been published in five volumes as *British plant communities* (Rodwell, 1991-2000) and is widely used in Phase 2 surveys. Summaries of the NVC descriptions for grasslands and montane communities, mires and heaths, and woodlands are also available for use in the field (Cooper, 1997, Elkington *et al* 2001, Hall *et al* 2001). There have been a number of changes to the classification since the draft version available when this handbook was written, necessitating revision of Appendix 8, which shows the relationship between Phase 1 categories and NVC communities.

The relationship between Phase 1 and the UK Biodiversity Action Plan Broad Habitat Classification can be found in Jackson (2000).

Additonal references for Appendices 8 and 9 only

Cooper, E A (1997) *Summary descriptions of National Vegetation Classification grassland montane communities.* UK Nature Conservation No. 14. Joint Nature Conservation Committee, Peterborough

Elkington, T, Dayton, N, Jackson, D L and Strachan, I M (2001) *National Vegetation Classification: field guide to mires and heaths.* Joint Nature Conservation Committee, Peterborough *www.jncc.gov.uk/communications/pubcat/heathland.htm.*

Hall, J E, Kirby, K J and Whitbread, A M (2001) *National Vegetation Classification: field guide to woodland.* Joint Nature Conservation Committee, Peterborough

Jackson DL (2000) Guidance on the interpretation of the Biodiversity Broad Habitat Classification (terrestrial and freshwater types): Definitions and the relationship with other habitat classifications. *JNCC Report, No. 307*
www.jncc.gov.uk/communications/pubcat/jnccreport.htm.

Rodwell, J S (ed.) (1991a) *British Plant Communities, Vol. 1: Woodlands and Scrub.* Cambridge University Press, Cambridge

Rodwell, J S (ed.) (1991b) *British Plant Communities, Vol. 2: Mires and Heaths.* Cambridge University Press, Cambridge

Rodwell, J S (ed.) (1992) *British Plant Communities, Vol. 3: Grasslands and Montane Communities.* Cambridge University Press, Cambridge

Rodwell, J S (ed.) (1995) *British Plant Communities, Vol. 4: Aquatic Communities, Swamps and Tall-herb Fens.* Cambridge University Press, Cambridge

Rodwell, J S (ed.) (2000) *British Plant Communities, Vol. 5: Maritime communities and vegetation of open habitats.* Cambridge University Press, Cambridge

Stace, C (1997) *New Flora of the British Isles,* 2nd ed. Cambridge University Press, Cambridge

Stace, C (1999) *Field Flora of the British Isles.* Cambridge University Press, Cambridge

CPSIA information can be obtained at www.ICGtesting.com
Printed in the USA
BVOW06s1347011115

424964BV00004B/6/P

9 781907 807244